Microservices in Action

Architecting Modular Systems

Table of Contents

Chapter 1. Introduction

Within the digital realm, the paradigm through which organizations develop, deploy, and maintain their software systems is constantly evolving. Diving into one of the most revolutionary trends in software architecture, our Special Report entitled, "Microservices in Action: Architecting Modular Systems," aims to break down this complex concept into layman's terms. Aided by real-life cases and expert commentary, this report elucidates how microservices can modularize systems, improve scalability, and streamline service continuity. Whether you're a seasoned tech professional, a budding software engineer, or a business leader eager to understand your technical team's strategies, this report presents practical insights into the adaptable world of microservices. Don't miss this opportunity to demystify the technical jargon and gain a comprehensive understanding of the role microservices play in reshaping the software landscape.

Chapter 2. Understanding the Fundamentals of Microservices

Microservices, also known as the microservice architecture, is an architectural style that structures an application as a collection of services that are highly maintainable and testable, independently deployable, organized around business capabilities, and can be owned by a small team. The microservice architecture enables the continuous delivery/deployment of large, complex applications. This architecture style represents a revolution in software design and deployment by enabling systems to be built as individual components that can function independently of each other.

Before we delve deeper, let's first picture the concept of software architecture.

2.1. The Concept of Software Architecture

Software architecture refers to the fundamental structure of software systems, constituting software elements, the externally visible properties of those elements, and their relationships. Software architecture deals with abstraction, with decomposition and composition, with style and aesthetics.

Just like a building, a software also has architecture. This architecture usually consists of various components/services that have been designed to function together to result in a software system working as a whole. These multi-service software are complex structures built to achieve a single, comprehensive task.

Among these multi-service software, two primary structural styles reign supreme: the monolithic architecture and the microservices architecture.

2.2. Transitioning from Monolithic to Microservices

Monolithic architecture, the traditional model for software systems, is comparable to a 'big ball of mud' — a singular, composite, and interlaced structure. Each component of the system is interconnected with the others — any alteration in one component will impact all others.

Monolithic systems have certain advantages — they are simple to develop, test, and deploy. However, as the scale and complexity of these systems grow, the limitations become glaring. A single codebase can become unwieldy. Scaling requires multiplying the entire codebase, not just the parts under load. Also, if an error occurs in one place, it can cause the entire system to crash.

The Microservice architecture, in contrast, counters most of these issues. Microservices take the approach of breaking down the tasks of a software system into separate services, each with a specific business focus. They advocate for the development and implementation of small services, with each service having its own process and communicating through mechanisms such as an HTTP/REST with resource APIs.

So, while monolithic applications are like single-family houses, microservices are like apartment units in a large building. They are independent - having their own utilities, entrance, and freedom.

2.3. Key Principles of Microservices

Microservices have certain foundational principles:

- Single Responsibility: Each microservice should have a single job. This facilitates the division of tasks and simplifies the development process.

- Independence: Microservices must be able to function independently. They should be self-reliant services, each with its own database and resources.

- Cross-functional Teams: Teams developing microservices should be cross-functional and include all the skills necessary for the development cycle.

- Infrastructure Automation: Deployment and scaling of microservices should be automated.

- Decentralized Governance: Each team should have the flexibility and independence to choose the best technologies and methods for their service.

- Failure Isolation: Failure in a single service should not impact others.

- Evolutionary Design: Enables systems to evolve over time. It supports continuous integration and deployment.

2.4. Benefits and Drawbacks of Microservices

The major benefits of microservices are their modularity and scalability. They increase the speed and productivity of software development processes because individual components can be developed, tested, and deployed independently. They also provide better fault isolation - a failure in one component doesn't impact the others.

However, there are drawbacks, the primary one being the complexity of managing multiple services compared to a single monolithic service. Handling inter-service communication can also be challenging. Also, distributed deployment might lead to data consistency issues.

In conclusion, microservices represent a way to avoid the limitations of traditional monolithic architecture, providing developers with a high degree of flexibility and independence. Its potential benefits clearly illustrate why it's garnering attention, and why you should consider it for your future software projects. However, the decision to use a microservice-based architecture should be based on the needs and context of your project, as well as your team's ability to confront and manage the challenges that come with it.

In the upcoming sections, we'll delve into more details of the microservice architecture, discussing design patterns, best practices, and case studies. By the end of this report, you should have a comprehensive understanding of microservices and will be able to make informed decisions when devising a software architecture strategy. It's an exciting journey ahead. Stay tuned!

Chapter 3. Traditional vs. Microservices Architecture: A Comparative Study

In the evolving world of software development, there is a growing shift from traditional monolithic architecture to the use of microservices. This shift is revolutionizing the way we build, deploy, and maintain software systems. In this comparative study, we will delve into the two contrasting architectural styles - Monolithic and Microservices - to understand their respective pros, cons, and appropriate use-cases.

3.1. The Monolith: Understanding The Traditional Approach

The traditional monolithic architecture designs all components of a system as one large unit. This unit includes everything: the user interface, backend server, databases, etc. However, maintaining this structure can become a nightmare with increasing system complexity. Its disadvantages include long upgrade cycles, less flexibility in adopting new technologies, and potential failures that might halt entire systems, to name a few.

For small systems, a monolithic approach benefits from simplicity in deployment and cross-component communication. It's easier to test, as everything exists in the same process. With low overhead, database transactions and operations exhibit high speeds. Yet as the system grows in complexity and size, these advantages diminish.

3.2. Paradigm Shift: The Microservices Approach

Microservices architecture emerged as an alternative, solving many challenges posed by the monolithic approach. In this architectural pattern, applications are divided into smaller, independently deployed units or services, each having its own exclusive domain and functionality. These microservices communicate with each other through APIs, often using HTTP/REST or asynchronous messaging.

Apart from easy deployment and maintenance, this approach fosters team autonomy, as every service can be developed and deployed by a separate team. Technological diversity is another advantage. Teams can choose the technology stack that best suits their specific service, without having to stick to limitations imposed by a broader architecture.

Furthermore, as each microservice can be modified or upgraded independently, system uptime has significant improvements. Only the updated service needs redeployment, allowing the other components to function as usual.

3.3. Comparative Case Analysis

Comparing monolithic and microservices architecture, let's consider a scenario of an e-commerce site scaling up its operations.

In a monolithic architecture, new features like "product recommendation" or "customer review" would require changes to the whole system, leading to increased complexity and longer deployment time. If any failure occurs during deployment, the entire system could potentially go offline.

Contrarily, in microservices architecture, each feature is a separate service. The introduction of a new feature or update doesn't affect other services. The "product recommendation" feature could fail without affecting the site's main functionality, presuming correct error handling. Although it's challenging to maintain data consistency and handle cross-service communication, solutions like incremental commit and event sourcing can help.

3.4. Trade-offs and Considerations

Moving to microservices doesn't entirely erase complexities; rather, it shifts them around. Microservices bring in challenges of service orchestration, inter-service communication, and subtleties of network latency. Moreover, versioning microservices can become complex if not managed correctly.

The choice between monolith and microservices isn't purely black or white; many grey areas surface depending upon system requirements. It's crucial to consider factors like system complexity, team size, expected growth, and scalability. When starting a new project, it's often recommended to start monolithic, gradually splitting into microservices as the system expands. This strategy mitigates risks associated with premature optimization.

Ultimately, the decision lies in assessing the trade-offs: the simplicity of a single unit in Monolithic versus flexibility and resilience of individual units in Microservices architecture.

3.5. Conclusion

The landscape of software architecture continues to evolve. Microservices architecture offers flexibility, scalability, and the ability to accommodate rapid changes, unlike the monolithic approach. However, it comes with its own set of challenges and considerations. Determining the right architectural approach primarily depends on your system requirement, business goals, and resource capacity. Broad exposure to and understanding of both monolithic and microservices architectures, along with an awareness of the trade-offs, is key for anyone interested in software engineering and technological strategy.

Chapter 4. Designing Microservices: Key Principles and Practices

For an effective and efficient Microservice architecture, proper design principles and practices are crucial. A wholesome approach to this architecture style revolves around considering certain key concepts, which we'll delve into in more depth throughout this section.

4.1. Understanding Domain-Driven Design

Domain-Driven Design (DDD) forms the foundation of effective microservice architecture. It involves modeling your microservices based on your problem domain, which is the business or organization area you're working on.

DDD proposes some principles and patterns that help you to design your microservices according to your business needs. The first step is to understand the business domain, followed by designing the software based on the complexity found in that realm.

The DDD modeling approach promotes the creation of a rich, versatile, and communicative model that captures the crux of the problem, which is then used to design the software. Businesses and development teams can then communicate using a unified language, reducing confusion and minimizing errors.

If you have a large system, it's best to break it down into various subdomains. Each subdomain is then encapsulated into its own model known as a Bounded Context. An organization's different departments can be visualized as separate bounded contexts. This way, each department (microservice) retains its model, with clear boundary lines segregating them.

4.2. Single Responsibility Principle

The Single Responsibility Principle (SRP) stipulates that a microservice should do one thing and do it well. One microservice must focus on a single business capability.

Segregating your system into different bounded contexts while abiding by the SRP means that when a change in business requirement occurs, it will affect only one or a few microservices, which results in minimized disruptions and improved development speed.

4.3. Polyglot Programming and Persistence

The idea behind Polyglot Programming is to use the best language, tools, and databases suitable for each microservice. New technologies can be tested with minimal risk because if they fail, the impact will be confined to a single microservice rather than affecting the entire system.

Polyglot Persistence, on the other hand, means choosing the right database for the job instead of using a single database type for all microservices.

The percentage of services that will use a Polyglot approach is generally low. However, having the flexibility to do so when necessary is a key factor of microservice architectures.

4.4. Isolation and Autonomy

A microservice needs to be fully self-contained and isolated from others. It should be autonomous, managing its own data, and be able to function independently from other microservices. Such segregation ensures that a failure in one service does not domino through the entire system, enhancing resilience and responsiveness.

This separation is not restricted to runtime isolation but extends to development time as well. Different teams should be able to develop, deploy, scale, and update their microservices independently of others.

4.5. Communication Styles

Synchronous communication happens in real-time like an API call requiring immediate response. Conversely, asynchronous communication doesn't require immediate response, for instance, using message queues or event-based communication.

Selecting the right communication style depends on your application's needs. Asynchronous communication is recommended whenever possible to avoid the ripple effect of latency and failure across the system.

4.6. Applying the Twelve-Factor App Methodology

The Twelve-Factor App methodology offers a structured way for building software-as-a-service apps that use declarative formats, have a clean contract with the operating systems, minimizing divergence between development and production, while ensuring scalability and ease of deployment. These factors are especially important considering the complexity and distributed nature of microservice architectures.

4.7. Continuous Delivery

Continuous delivery, an integral part of agile software development, establishes a streamlined, automated route for taking software from design to deployment. By utilizing small, incremental changes, software is consistently ready for deployment, enhancing reliability, and improving development and production environment synchronicity.

4.8. ...and Last but not Least, Embrace Failure

Microservices should embrace failure and develop strategies to handle them effectively. Tools like circuit breakers, service meshes, and retry pattern, as well as concepts like bulkheads, will help to build resilience in your microservice architecture.

These concepts encompass the design principles and practices pivotal to microservice architecture. By understanding and applying these, organizations can better navigate the complex realm of microservice design, creating robust and efficient systems that meet their unique business needs and goals.

The adoption of these principles isn't a panacea, nor does it guarantee system perfection. But they cultivate an environment that fosters better understanding, planning, developing, and maintaining microservices in an effective manner.

Chapter 5. The Role of APIs in Microservices Architecture

A vital facilitator in the world of microservices architecture is the Application Programming Interface, or API. APIs sit at the nucleus of microservices, making the separated services communicable. Stated simply, APIs provide a set of rules and protocols for building and integrating application software. To explore this in greater detail, let's delve into the role APIs play in microservices architecture.

5.1. API - A Bridge Among Microservices

APIs serve as the connectors allowing discrete microservices to converse, exchange data, and orchestrate processes. Laying down the rules for interaction, APIs describe what's possible and the constraints on how one service can communicate with another. They are similar to the concept of an electrical socket, establishing a standard way for devices to plug into a power supply. Without the socket, the device can't access the electricity

5.2. APIs and Service Independence

Despite being closely linked via APIs, each microservice is, by design, an independent entity. It provides a well-defined function and has its own dedicated database, reducing dependencies and allowing each service to evolve individually, independently, and at its own pace. While microservices talk to each other, each one remains agnostic to the inner workings of the rest. Ideally, a change, replacement, or update in one service should not impact others. APIs, by forming a standard protocol for communication and data exchange, strike this tricky balance of interdependence and independence.

5.3. Approaching API Design in Microservices

Creating an efficient API is pivotal for the successful implementation of a microservices architecture. A well-structured API can make services easier to manage and debug, and more scalable. However, good API design is not just about defining how software components interact but also managing evolution and change. There are several approaches, but three of the most common are the REST API, gRPC, and GraphQL.

REST (Representational State Transfer), an architectural style rather than a protocol or standard, remains popular among developers due to its simplicity and suitability for web services. REST APIs use simple HTTP protocols and JSON for data transfer, greatly enhancing compatibility, interoperability, and simplicity.

gRPC is a high-performance, open-source framework developed by Google. It uses Protocol Buffers (protobuf) as its interface definition language, enabling definition of services and message types in a single .proto file.

GraphQL, on the other hand, provides a querying language specifically built for APIs, offering more power and flexibility than REST. Rather than working with fixed data structures, clients can request exact data as needed, reducing over-fetching and under-fetching problems.

5.4. API Gateway in Microservice Architecture

A key component of any microservice-based architecture, an API Gateway, often plays the role of a reverse proxy, routing requests from clients to appropriate services. The Gateway hides the complexity of underlying microservices, thereby mitigating the challenges of microservices communication.

It's crucial to note that, in a microservice architecture, microservices often need to communicate with one another. These communications tend to be either "synchronous" or "asynchronous". APIs support synchronous communication, where the caller waits for the callee's response. Asynchronous communication, where the caller does not wait for a response (often implemented with message queues), is also important but falls outside the API's domain.

5.5. Versioning and Microservices

Versioning is a cardinal part of maintaining APIs in microservices architecture. Due to the decoupled nature of services, when an API changes, it should not impact others. To prevent this, developers implement version control.

API versioning can be handled in multiple ways, such as URL path, request header, or query parameters. While versioning is hotly debated, the overall goal remains - to prevent breaking changes and maintain a well-documented, efficient API.

In conclusion, APIs form the crux of communication in microservices architecture. They establish a standard protocol for communication, facilitating necessary independence and successful integration. Understanding them and investing in their design, development, and maintenance is key to unlocking the benefits of microservices.

Chapter 6. Scaling and Deploying Applications Using Microservices

The ability to scale and deploy applications effectively is a quintessential requirement in this digital era. With the evolution of software development methodologies, microservices have revolutionized how applications are structured, developed, and rolled out by enabling a highly scalable and fault-tolerant architecture.

6.1. Understanding the Scaling Needs

An essential aspect of microservices deployment strategy is knowing *when* and *how* to scale your services. Two types of scaling options are available: vertical and horizontal scaling.

Vertical scaling, often referred to as "scaling up," involves increasing the capacity of a single server by adding more power (CPU, RAM). The limitation of this method is the constraints of a single machine.

Horizontal scaling, on the other hand, often termed as "scaling out," is achieved by adding more servers to the system to share the processing load. This method is more effective due to the distributed nature of processing.

In microservices architecture, services are granular and independently deployable, allowing for more focused horizontal scaling, i.e., you scale only those services that are experiencing high consumption demands.

6.2. Implementing Scalability with Microservices

Microservices support both vertical and horizontal scaling as you have the flexibility to use a combination of both, depending on your project needs.

To implement scaling efficiently:

- Ensure services are stateless: Stateless microservices are those which do not store any client data from one session to the next. This allows any subsequent requests to be handled by any available microservice instance.

- Use a load balancer: A load balancer is a tool that distributes network or application traffic across a cluster of servers. It helps to increase concurrent user capacity and reliability of your applications.

- Implement Service Discovery: Service discovery is crucial in a dynamic environment shaped by the continuous building and tearing down of microservices.

6.3. Deploying Microservices

Microservices deployment comes with the challenge of managing multiple, independently deployable components of an application. Some strategies include:

- Containers: Containers are a lightweight, stand-alone, and executable software package that includes everything needed to run a piece of software. Docker is one of the most popular container platforms used today.

- Orchestrator: Managing a large number of containers and services could be a daunting task. Orchestrators like Kubernetes can help to automate the deployment, scaling, and management of containerized applications.

- Continuous Deployment: Microservices enable continuous deployment where changes in the software are automatically bug-checked and uploaded to the repository. Tools like Jenkins can help in achieving continuous deployment.

6.4. Managing Data Consistency

When scaling with microservices, managing data consistency becomes a significant challenge as each service has its own database. This leads to difficulties in transactions that span multiple services. Strategies for managing distributed data include:

- Implementing Sagas: A saga is a sequence of local transactions where each transaction updates data within a single service. Sagas are essential to ensure data consistency among different microservices.

- Using Domain-Driven Design: Domain-driven design (DDD) focuses on the core domain and domain logic, improving collaboration between technical and domain experts, achieving data consistency.

6.5. Ensuring Service Continuity

Achieving service continuity in a microservices architecture is another challenge as it involves having a fault-tolerant system in place. This can be achieved by:

- Setting up Health Checks: Health checks help in monitoring the statuses of individual microservices and ensuring they're running optimally.

- Implementing Circuit Breakers: Circuit breakers prevent a network or service failure from cascading to other services.

- Leveraging the Sidecar Pattern: The sidecar pattern is a single-node pattern consisting of an application (parent) and an attached sidecar service, where the parent application and the sidecar share a lifecycle.

Scaling and deploying applications using a microservices architecture, while challenging, reaps numerous benefits such as enhanced scalability, improved performance, and the ability to ensure service continuity. With the right strategies and tools in place, businesses can significantly enhance their operational efficiency and agility in the digital landscape.

Chapter 7. Case Study: Microservices in Large Scale Enterprises

In the digital age, the approach to adopting microservices in large-scale enterprises is a strategic decision that's made around scale, speed, resilience, and team autonomy. This chapter will decisively charter this journey by delving into the architecture, challenges, benefits, and implementation techniques of microservices in well-established companies.

7.1. The Journey to Microservice Architecture

The starting point of many legacy enterprises has typically been a monolithic structure in which the software's components are interconnected and interdependent, leading to unwieldy software blobs that are hard to manage and maintain. To mitigate this, large-scale enterprises have migrated to adopt microservice architecture, a style that structures applications as a collection of loosely coupled, independently deployable services.

Implementing microservices in large-scale enterprises involves a transition from a single, unified system to discretely managed and independently operating components. This modular approach allows an application to be broken down into smaller, manageable parts that can be independently developed, maintained, and scaled, thus providing increased flexibility and efficiency.

7.2. The Microservices Implementation Process

The transition to microservices is a substantial undertaking that involves a marked shift in development techniques, team structure, and operational paradigms. One of the first steps during this shift includes identifying business capabilities and translating them to appropriate services. The services should be designed with loose coupling and high cohesion in mind, allowing for future modification and reusability.

Automating the deployment process is of critical importance when adopting a microservices approach. Deployment should be repeatable and reliable, ideally with little to no human intervention. Furthermore, the release process should be structured in such a way that if a component fails, it does not cause system-wide outage.

Continuous integration and continuous delivery (CI/CD) pipelines can play a pivotal role in minimizing development-to-production lead time, ensuring system reliability, and providing frequent updates and feedback. With the help of automation tools, large-scale enterprises can streamline these processes and adapt quickly to changing business requirements.

Testing microservices can present unique challenges; however, using strategies like service virtualization, contract testing, and end-to-end testing can ensure reliable, well-functioning services.

7.3. Challenges and Solutions

Scaling to a large number of services presents a complex array of challenges, such as service coordination, data consistency, network latency and fault tolerance. It's essential to understand the trade-offs and carefully design the system to mitigate these constraints.

Additionally, the complexity of managing a large number of services and the associated inter-service communication can be daunting. This can be addressed by keeping the services size small, adopting standards and conventions, using API gateways and implementing service discovery.

Another challenge that large enterprises could face is ensuring data consistency across services. Since each service manages its own database in a microservices architecture, the issue of maintaining data consistency arises. This is typically addressed by implementing a distributed transaction system or relying on eventual consistency, where a compromise is made to prioritize availability and partition tolerance.

7.4. Real-Life Cases

Several multinational corporations have successfully transitioned to microservices, reaping the numerous benefits this architecture style offers. Companies such as Netflix, Amazon, and eBay have adopted the approach, decreasing their time to market, scaling operations, and improving customer experience.

For example, Netflix's transition from a monolithic structure to a microservices architecture allowed the company to scale to support over a hundred million global customers. The architecture gave Netflix the flexibility to deploy thousands of changes per day and perform A/B testing to learn and iterate rapidly.

Microservices have also proven beneficial for large-scale financial institutions. Capital One's adoption of this architecture has enabled them to go from deploying twice a week to over fifty deployments per day, considerably enhancing their speed of innovation.

7.5. A Glimpse into the Future of Microservices

Considering these cases and the benefits, it's increasingly clear why large companies are moving towards microservices and making it core to their strategy. Its role in facilitating agility, resilience, and speed of development will continue to make it a compelling architectural style for the foreseeable future.

In summary, transitioning to a microservices architecture in a large-scale enterprise presents a unique set of challenges and benefits. Through strategic planning, organizational readiness, and targeted tooling, these challenges can be overcome, leading to increased agility, fault tolerance, and increased deployment velocity. The journey requires a significant commitment, a shift in culture, and a willingness to embrace change. However, the potential benefits and opportunities for innovation make it a worthwhile pursuit.

Chapter 8. Overcoming Challenges and Pitfalls in Microservices Implementation

Understanding and effectively navigating through the challenges and potential pitfalls inherent to microservices implementation is pivotal for any organization seeking to modernize their software architecture. This can range from determining and shaping microservice boundaries, providing adequate security measures, or handling inter-service communication effectively. Let's delve deeper into these challenges, explore potential pitfalls, and provide best practices to combat these.

8.1. Identifying and Shaping Microservice Boundaries

One of the foremost obstacles in microservices implementation lies in accurately identifying and shaping microservice boundaries. Doing so can be an intricate process that necessitates clear understanding of each business functionality, and the inter-relationships between them.

Business capabilities and their corresponding data entities should dictate the creation and encapsulation of microservices. A common pitfall here is over or under partitioning, which can not only result in convoluted service calls, but also impede subsequent maintenance and enhancement of these services. Avoiding this requires due diligence in understanding business context and requirements, ensuring that each microservice has its own bounded context and that it aligns with a single business capability.

8.2. Securing Microservices

Microservices architecture intrinsically tends to increase attack surfaces due to distributed deployment and communication between services. Failure to adequately secure each individual service can escalate the potential for security breaches.

Safeguarding microservices involves ensuring appropriate authentication, authorization, and integrity checks. Using an API Gateway for managing secure entry points into the system, applying role-based access control, and implementing encryption for data at rest and in transit are some recommended practices. Additionally, defense-in-depth techniques can be utilized to provide multiple security layers and reduce the likelihood of unauthorized system access.

Tools like OAuth and JWT can be used for securing access to APIs, and service mesh solutions such as Istio or Linkerd provide capabilities like mutual TLS to secure inter-service communication.

8.3. Managing Inter-Service Communication

In a distributed architecture like microservices, managing the communication between multiple services is a prominent stumbling block. The asynchronous character of microservices communication combined with inevitable network failure can result in complex error handling scenarios.

Using a suitable communication protocol is key. While REST might be sufficient for synchronous, simpler scenarios, asynchronous protocols like AMQP or MQTT can be adopted for more complex, event-driven designs.

Circuit breaker patterns can be implemented to safeguard services from cascading failures and system-wide outages. Tools such as Hystrix and resilience4j offer built-in circuit breaker functionalities. A message queue, such as RabbitMQ or Apache Kafka, can also be utilized to provide reliable, contract-based messaging between services.

8.4. Handling Data Consistency

Microservices tend towards decentralized data management, which pose challenges in ensuring data consistency. Unlike monolithic systems where ACID properties ensure transactional consistency, microservices typically rely on BASE (Basically Available, Soft state, Eventually consistent) principles.

Implementing eventual consistency introduces complexities related to replication, propagation of updates, and resolution of conflicts. Patterns such as Saga can be used to manage data consistency across services. Events can be used to signal state changes, followed by compensating operations in case of failures.

8.5. Operational Overhead

Microservices introduce an operational overhead due to their number, granularity, and distributed nature. Monitoring, logging, and debugging across many independent services can be challenging, and standardizing these becomes vital.

Conversely, deploying microservices in containers can assist in standardizing environments and streamline deployment. Container orchestration tools such as Kubernetes help manage containerized applications at scale.

For maintaining application health visibility, centralized logging and monitoring solutions like the ELK stack (Elasticsearch, Logstash, Kibana) or Grafana can be deployed. Tracing tools like Zipkin or

Jaeger can aid in troubleshooting distributed systems.

8.6. Organizational Challenges

Microservices not only bring about a shift in technology but also in the organizational structure and culture. Adopting a microservice architecture often requires transforming traditional project teams into cross-functional units or squads, an approach known as the "Conway's Law".

While cross-functional teams encourage autonomy and faster decision-making, they also impose challenges like aligning multiple teams to the overarching business goals and managing dependencies. Formulating well-defined guidelines, clear contracts between teams, leadership support, and effective communication can manage these challenges.

In conclusion, optimizing a microservice architecture demands a deep understanding of business requirements, careful planning, proper tool selection, and an alignment of technological choices with organizational culture. By adequately addressing the outlined pitfalls and challenges, an organization can reap the benefits of microservices-- increased scalability, flexibility, and a resilient system architecture.

Overall, the journey to effectively implement microservices may be fraught with challenges, but the path becomes less treacherous when armed with the right knowledge, tools, and methodologies.

Chapter 9. Microservices and the Cloud: A Symbiotic Relationship

Microservices architecture has consistently gained traction in the domain of software development for its ability to provide seamless scalability and enhanced service continuity, resulting in modernized and modular systems. A significant, yet often overlooked, component that enhances the efficacy of microservices methodology is cloud computing. The synthesis of microservices and cloud technology engenders a symbiotic relationship that is fast becoming the backbone of digital infrastructures worldwide.

9.1. Understanding the Relationship

Before delving into the intricacies of how microservices and the cloud interrelate, it is paramount to comprehend what these individual components entail. Microservices, often termed as microservices architecture, refers to a style of structuring applications as collections of loosely coupled, independently deployable services, each running unique processes and communicating over networked APIs. This ensures better modularity, making applications easier to scale and develop, thus, aiding faster innovation and potential risk mitigation.

On the other hand, cloud technology is a model that enables ubiquitous access to shared pools of configurable computing resources. In simpler terms, it's a platform providing on-demand services via the "Internet" with the capability to scale up or down according to the workload volume, effectively reducing the cost while increasing the scalability and availability of applications.

The interplay of microservices architecture and cloud technology creates an asynchronous, loose coupling, enabling tasks to be developed, deployed, and scaled independently. This symbiosis proffers many potentiating benefits that this chapter cares to illuminate.

9.2. Benefits of Microservices Architecture and Cloud Technology Cooperation

Microservices architecture, when supported by cloud technology, fashions a plethora of benefits that collectively streamline application management and bolster service continuity. Here is an overview of some of these benefits:

- Improved Scalability: The independent scaling of each microservice becomes easier and more efficient in the cloud. You simply scale out the resource (CPU, memory, etc.) guzzling services without necessarily scaling out the whole application. This functionality is one of the cornerstones of effective resource utilization.

- Enhanced Resilience: Isolating services reduces the likelihood of an application-wide failure. When one service fails, it doesn't hold back other services or cause the entire application to crumble. In conjunction with cloud technology, these isolated services can be easily restarted or substituted.

- Faster time-to-market: The cloud offers a myriad of tools and services to assist in the continuous integration and deployment of microservices. This approach decreases the time required to effectively prototype, test, and release new application iterations, enabling a shorter go-to-market timeframe.

- Optimal resource usage: Allocating different microservices to differently sized cloud-based servers can help optimize resources as compared to traditional monolithic architectures. Each microservice consumes resources in proportion to the demands of its function only.

- Facilitated process optimization: The pairing of monitoring and analytics tools with the distributed nature of microservices can lead to improved and in-depth insights about performance and user behavior.

9.3. Real-World Scenarios Demonstrating Microservices and Cloud Interplay

A symbiotic relationship between microservices and the cloud translates to digital solutions that are scalable, resilient, and efficient. Let's delve into the practical implementation of this amalgamation, referencing real-world use cases.

- Netflix: This streaming giant adopted a cloud-based microservices architecture to cater to its growing global customer base. By splitting up their monolithic application into hundreds of smaller microservices, each performing a specific function, Netflix became capable of deploying, scaling, and updating each one independently, successfully meeting fluctuating demand and ensuring smooth customer experiences.

- Amazon: Amazon adopted a microservices architecture on a cloud platform to manage traffic and enhance scalability. Their architecture includes numerous small services, each responsible for delivering a specific functional benefit. This flexibility allows Amazon to optimize and update services individually without impacting the overall application, maintaining optimal performance during traffic surges.

- Uber: To scale rapidly worldwide, Uber transitioned from a monolithic architecture to a microservices approach utilizing cloud technology. Their services include trip, payment, and driver-location management, each with its unique scaling and deployment requirements. By separating these services, Uber can scale efficiently without overburdening resources.

9.4. Conclusion

A seamless blend of microservices architecture with cloud technology offers unparalleled benefits, enabling modern businesses to maintain agile, scalable, and resilient applications. The successful integrations by industry titans like Netflix, Uber, and Amazon are testament to the value proposition of the symbiotic relationship.

While the coupling of microservices with the cloud simplifies application development and management, implementing this amalgamation effectively requires deliberate planning and continuous refinement with an understanding of the organization's needs and capabilities. The advancements in this space are demonstrating that the interplay between cloud and microservices is not a trend; it's the future of software architecture.

Chapter 10. The Future of Microservices: Predictions and Possibilities

With the rise and evolution of microservices within the industrial paradigm, it has become evident that there's no question regarding their potential and significance. Many organizations are increasingly shifting from monoliths to adopt microservices as their preferred architectural style. However, as we stand on the brink of a technology era built on a microservices foundation, it's vital to understand where we might be headed and how innovations in this sector could continue shaping the future. In this chapter, we will delve into an in-depth analysis of future predictions and possibilities involving microservices, examining both the technological advancements likely to revolutionize this sector and the potential challenges that may ensue.

10.1. The Technological Landscape

As with any technical domain, the advancements of microservices are not in isolation. Several complementary technologies could drive the surge forward or add another dimension to how microservices are used. Technologies like serverless computing, Service Mesh, artificial intelligence (AI), and Kubernetes are among the potential game-changers.

Serverless computing, for example, eliminates the need to manage infrastructure. This hands-off approach reduces the overhead and complexity often associated with deploying microservices. The popularity of serverless computing will likely continue to influence the development and deployment of microservices.

AI can be another powerful ally for microservices, especially when it comes to maintaining and managing large-scale microservices ecosystems. Future AI-based systems could potentially rectify issues proactively, automate repetitive tasks, and optimize resource allocation among the deployed microservices.

Service Mesh, a configurable infrastructure layer for a microservices application, will facilitate service-to-service communication while also providing crucial capabilities like service discovery, load balancing, failure recovery, metrics, and monitoring. This infrastructure is likely to be more ubiquitous as microservices become more complex and interconnected.

Kubernetes, an open-source system for automating deployment, scaling, and management of application containers, has already played a crucial role in getting microservices to where they are today. We predict a more significant role for Kubernetes in the future, given its benefits and wide adoption.

10.2. Expanded Use Across Industries

Microservices, while popular across several industries like banking, e-commerce, and health care, are likely to find expanded use in other areas not traditionally associated with this technology. This widespread adoption might be motivated by the agility and scalability benefits that a microservices architecture can offer.

Even within industries already adopting microservices, we forecast that usage will become more nuanced and sophisticated, leading to bespoke microservices solutions catering to the specific demands of different industry sectors.

10.3. User Experience: The Pivotal Component

The application of microservices will emphasize the importance of user experience in the future, with a growing focus on creating multiple front-ends. This way, the front-ends can interact with different subset services in service-oriented architecture (SOA), delivering an improved, personalized user experience.

10.4. Data Management: Challenges and Solutions

As we foresee the proliferation of microservices in the future, handling data will be a significant challenge. With thousands of microservices potentially needing to access data simultaneously, data management solutions must evolve to allow distributed transactions and handle concurrency conflicts.

Solutions such as event sourcing or Command Query Responsibility Segregation (CQRS) will become crucial tools to address these data management complexities, ensuring that microservices will continue to work efficiently and securely, even when handling vast and intricate data systems.

10.5. Security and Compliance

While microservices can bring numerous benefits, they also surface new security challenges given that potential attack surfaces are multiplied considerably. Future developments will need to prioritize creating more robust security architectures and incorporating compliance mechanisms within the microservice design.

The prospects for microservices are bright and varied. As we steer into the future, the confluence of microservices with other advanced technologies—with AI and serverless architecture leading the charge—offers fascinating possibilities. The challenges, while substantial, are not insurmountable and should drive innovative solutions. Ultimately, the microservices architecture is poised to remain at the forefront of organizational digital strategies, transforming operational effectiveness and customer experience.

Chapter 11. Building a Business Case for Microservices: Cost-Benefits Analysis

The decision to migrate from a monolithic system to microservices is not one to be taken lightly. It involves a considerable shift in organizational thinking, widespread process changes, and a significant investment in time and resources. Consequently, there is an essential need for a quantifiable business case that justifies this investment by identifying clear, tangible benefits. This chapter will closely examine the benefits of adopting a microservices architecture, including a thorough cost-benefits analysis. It is important to note, though, that while this analysis can provide valuable insight, actual outcomes will vary based on the unique circumstances of each organization.

11.1. Understanding the Investment

Before we delve into the advantages of microservices, it is crucial to grasp a fundamental understanding of what the transition entails. Implementing microservices architecture typically includes refactoring the existing monolithic application into smaller services and setting up communication between them. Additionally, there will be considerations for data management, creating APIs for each service, implementing service discovery, and setting up a process for deployment, scaling and monitoring each of the services.

Investment goes beyond merely financial expenses. It also includes effort and critical resources. It encompasses variables like time for planning and implementing the new system, training staff, disrupting regular business operations, possible downtime during the shift, and long-term requirements for maintenance and system evolution.

11.2. Cost Efficiency

Microservices deliver numerous cost benefits, primarily through higher efficiency and resource optimization. By breaking down applications into small, autonomous services, resource utilization is significantly improved. Services can then be independently deployed and scaled on an as-needed basis. If one microservice encounters high demand, it can be separately scaled without impacting other services or requiring more extensive infrastructure investments.

Moreover, the modularity of microservices encourages code reuse. This benefit can be particularly impactful in large organizations with many shared functions. Instead of repeatedly coding the same functions, developers can create a shared library of services that can be utilized across multiple projects, thereby saving time and reducing costs significantly.

11.3. Agility and Speed

The decoupled nature of microservices implies that each service can be developed, tested, and deployed independently. This ability facilitates agile development processes, enabling fast iterations and allowing businesses to respond quickly to changing market conditions. Instead of revamping an entire system for a new feature, teams can update a specific service, resulting in quicker turnaround times and improved time-to-market.

This increase in development speed does not just improve adaptability; it also significantly impacts overall costs. In today's fast-paced digital economy, speed to market can be a crucial competitive advantage, and being the first to introduce a new feature or service can lead to higher market shares, improved revenue streams, and stronger customer loyalty.

11.4. Resilience

Resilience means the system's ability to resist, respond to, and recover from faults. In a monolithic architecture, if one component doesn't function correctly, it can take down the whole application. But in the microservices architecture, since services are loosely coupled and operate independently, if a single service fails, the entire application doesn't necessarily need to be compromised. This improves system resilience, and less downtime means you're saving money in terms of remediation costs and potential lost revenues due to system outages.

11.5. Summary

While there are clear cost advantages of implementing microservices, businesses should consider their unique context before moving forward. A microservices architecture might bring considerable benefits for complex, large-scale applications that require high rates of change and scalability. Conversely, for small systems and simple applications, switching to microservices might be an overinvestment.

In any change of such significance, a mindful, data-informed decision is advisable. Analyze your own business needs, existing system architecture, team capabilities, and long-term strategic goals. Stakeholders should be engaged and ensured transparency. As a final caution, remember that the migration to a microservices architecture is a marathon, not a sprint. It should be approached as a long-term strategic decision rather than a patchwork solution.

www.ingramcontent.com/pod-product-compliance
Lightning Source LLC
Chambersburg PA
CBHW071047260726
48661CB00007B/3180